THE AMERICAN LANDSCAPE

EVERGLADES

DWAYNE HICKS

PowerKiDS press

Published in 2026 by The Rosen Publishing Group, Inc.
2544 Clinton Street, Buffalo, NY 14224

First Edition

Editor: Greg Roza
Book Design: Michael Flynn

Photo Credits: Cover, p. 1 Irina Wilhauk/Shutterstock.com; (series background) Omeris/Shutterstock.com; p. 5 Simon Dannhauer/Shutterstock.com; p. 7 Radzas2008/Shutterstock.com; p. 9 https://commons.wikimedia.org/wiki/File:Seminole_War_in_Everglades.jpg; p. 11 Helen89/Shutterstock.com; p. 13 ocudrone/Shutterstock.com; p. 15 (cypress) Anthony Ricci/Shutterstock.com; p. 15 (mangrove) Alejandro Espaillat/Shutterstock.com; p. 17 (Florida alligator) SunflowerMomma/Shutterstock.com; p. 17 (flamingo) Cris D/Shutterstock.com; p. 17 (manatee) Harry Collins Photography/Shutterstock.com; p. 19 Heiko Kiera/Shutterstock.com; p. 21 Maridav/Shutterstock.com.

Library of Congress Cataloging-in-Publication Data

Names: Hicks, Dwayne, author.
Title: Everglades / Dwayne Hicks.
Description: [Buffalo, New York] : PowerKids Press, [2025] | Series: The American landscape | Includes bibliographical references and index.
Identifiers: LCCN 2024034280 (print) | LCCN 2024034281 (ebook) | ISBN 9781499450897 (library binding) | ISBN 9781499450880 (paperback) | ISBN 9781499450903 (ebook)
Subjects: LCSH: Natural history–Florida–Everglades–Juvenile literature. | Wetland ecology–Florida–Everglades–Juvenile literature. | Everglades National Park (Fla.)–Juvenile literature. | Everglades National Park (Fla.)–Environmental conditions. | Everglades (Fla.)–Juvenile literature. | Everglades (Fla.)–Environmental conditions–Juvenile literature.
Classification: LCC QH105.F6 H53 2025 (print) | LCC QH105.F6 (ebook) | DDC 557.59/39–dc23/eng/20240731
LC record available at https://lccn.loc.gov/2024034280
LC ebook record available at https://lccn.loc.gov/2024034281

Manufactured in China

CPSIA Compliance Information: Batch #QSPK26. For Further Information contact Rosen Publishing at 1-800-237-9932.

CONTENTS

One-of-a-Kind Place

The Everglades is a one-of-a-kind place. Marked by flatland, grassland, wetland, and **estuaries**, the Everglades offers visitors a lot to see and do! This large part of Florida has many kinds of wild animals. Some are considered **invasive species**. Let's go to the Everglades!

Where Is It?

The Everglades covers most of southern Florida. It's hot and wet for most of the year. This land **slopes** toward the ocean. There are many lakes and rivers. Many roads cross this area, and you can find many towns and cities there as well.

Atlantic
Ocean
FLORIDA
Gulf of Mexico
Tampa
West
Palm
Beach
Lake
Okeechobee
Everglades
Miami
Everglades
National Park

The Everglades' History

Native Americans have lived in Florida for more than 14,000 years. Some lived along the west coast. They made waterways to travel by boat. They used the Everglades for food and water. Later, in the 1700s, the Seminole Indians used the Everglades to stay safe from European settlers.

Americans started hunting Everglades animals for their feathers and hides in the 1800s. They built waterways to send water of the Everglades to farms and communities. On December 6, 1947, the southernmost part of Florida became Everglades National Park. Today, the area is known for **tourism**.

EVERGLADES Safari

River of Grass

Water flows south from central Florida through rivers and lakes. Much of this water gathers in Lake Okeechobee. Water flows out of the lake, forming a slow river that can reach 60 miles (97 km) wide and 100 miles (161 km) long. This river flows through different kinds of grasslands.

Everglades Plants

Grass is one of the most common plants in the Everglades. Large areas of grass allow water to flow south. Cypress trees grow on land covered with water. Mangrove trees grow in rivers and estuaries. Their large roots keep them in the ground.

cypress
mangrove

Everglades Animals

The Everglades is home to many kinds of animals. These **include** the Florida alligator and the Florida panther. Many kinds of birds live there—flamingos, storks, pelicans, owls, herons, and more. You may even see a manatee! Other animals include turtles, otters, deer, and snakes.

Florida alligator

flamingo

manatee

You Don't Belong Here!

The Everglades is home to invasive species. Burmese pythons and other large snakes are a big problem. So are wild boars, a kind of pig, and the walking catfish—which can hop from one river to another! Invasive species make it hard for other plants and animals to live.

Human Actions

The flow of water through the Everglades naturally cleans the area. However, the actions of people have changed the flow of water. The land that the Everglades covers has become smaller. Parts of the Everglades are **polluted**. Many people are working hard to help restore, or fix, the Everglades.

GLOSSARY

estuary: A body of water where salt water and fresh water meet.

include: To take in or have as part of a whole or group.

invasive species: Plants and animals from one area that have been moved to a new area, where they can be harmful to other plants and animals.

pollute: To make dirty.

slope: To be higher on one end than the other.

tourism: The practice and business of traveling for fun.

FOR MORE INFORMATION

BOOKS

Jacobson, Bray. *The Everglades.* Buffalo, NY: Gareth Stevens, 2023.

Walton, Kathryn. *Visit Everglades National Park!* Buffalo, NY: Enslow Publishing, 2025.

WEBSITES

National Geographic Kids: Burmese Python
kids.nationalgeographic.com/animals/reptiles/facts/burmese-python
Read about the Burmese python, and see close-up photographs of them.

National Geographic Kids: Everglades National Park
kids.nationalgeographic.com/nature/article/everglades
Learn more about the Everglades at this informative website.

Publisher's note to educators and parents: Our editors have carefully reviewed these websites to ensure that they are suitable for students. Many websites change frequently, however, and we cannot guarantee that a site's future contents will continue to meet our high standards of quality and educational value. Be advised that students should be closely supervised whenever they access the internet.

INDEX